ANIMAL RHYME TIME!

RHYME TIME WITH SNAKES!

BY JONAS EDWARDS

Gareth Stevens PUBLISHING

Please visit our website, www.garethstevens.com. For a free color catalog of all our high-quality books, call toll free 1-800-542-2595 or fax 1-877-542-2596.

Library of Congress Cataloging-in-Publication Data

Names: Edwards, Jonas, author.
Title: Rhyme time with snakes! / Jonas Edwards.
Description: New York : Gareth Stevens Publishing, [2021] | Series: Animal rhyme time! | Includes index.
Identifiers: LCCN 2020004004 | ISBN 9781538256121 (library binding) | ISBN 9781538256107 (paperback) | ISBN 9781538256114 (6 Pack) | ISBN 9781538256138 (ebook)
Subjects: LCSH: Snakes–Juvenile literature. | Rhyme–Juvenile literature.
Classification: LCC QL666.O6 E39 2021 | DDC 597.96–dc23
LC record available at https://lccn.loc.gov/2020004004

First Edition

Published in 2021 by
Gareth Stevens Publishing
111 East 14th Street, Suite 349
New York, NY 10003

Editor: Kate Mikoley

Photo credits: Cover, p. 1 Vaclav Sebek/Shutterstock.com; cover, pp. 3–24 (music notes) StockSmartStart/Shutterstock.com; p. 5 Helen H. Richardson/The Denver Post via Getty Images; p. 7 kristianbell/ RooM/Getty Images; p. 9 Capt Suresh Sharma/Moment/Getty Images; p. 11 Ariadne Van Zandbergen/ Lonely Planet Images / Getty Images Plus; p. 13 liveslow/ iStock / Getty Images Plus; p. 15 George Grall/ National Geographic Image Collection/Getty Images; p. 17 Westend61/Getty Images; p. 19 John Cancalosi/ Photolibrary / Getty Images Plus; p. 21 Deki Hendrik / EyeEm/ Getty Images.

Printed in the United States of America

CPSIA compliance information: Batch #CS20GS: For further information contact Gareth Stevens, New York, New York at 1-800-542-2595.

CONTENTS

Boldface words appear in the glossary.

Watch for Snakes!

Ssssss. Listen. Do you hear
that sound?
It might mean that there's a
snake around!
Not all snakes want to harm you
or strike.
Still, watch out for snakes if you
go for a **hike**.

Slithering Snakes

Snakes are reptiles, a kind of animal covered in scales.

With no **limbs**, snake bodies sort of look like long tails.

Reptiles are **cold-blooded** and most lay eggs.

But how do snakes move if they don't have legs?

SCALES

Slithering is the sliding way snakes
get from place to place.
Some can move fast, while others
have a slower **pace**.
Flat scales called scutes on their
bellies help snakes move.
The more scutes a snake has, the
more its speed will **improve**.

SCUTES

Snake Snacks

Snakes are predators and eat living
animals called prey.
People chew food, but snakes don't
dine this way.
They swallow their food whole,
in one big bite!
The snack can be bigger than
them—what a sight!

Some snakes hunt, but others just wait.
By the time the prey sees them,
it's often too late!
Some snakes use **venom** and bite
to kill their prey.
Others **squeeze** until the animal can't
breathe or get away.

VENOM

Venom Can Cure?

Snake venom can kill, that's for sure.
But it's also been used in drugs
as a **cure**!
There are thousands of species, or
kinds, of snakes in the world.
You can tell one might be cold or
scared if its body is curled.

Snakes All Around

Snakes have been found on every
continent but one.
Antarctica is too cold for them to
have any fun!
Snakes make their homes in sea
and on land.
They can be found in places with
mountains, grass, and sand.

Slipping Off the Skin

Snakes shed, or get rid of, their skin
as they grow!
They rub things and wiggle out of
the skin as they go.
It comes off inside out like a sock.
It may get left in the grass or out
on a rock.

Give Snakes Space

If you see a snake, it's best to
stay away.
It's probably not interested in
having you as its prey.
Still, venomous snakes can bring
danger wherever they go.
But even venom can be helpful
sometimes, as you now know!

GLOSSARY

cold-blooded: having a body temperature that's the same as the temperature of the surroundings

continent: one of Earth's seven great landmasses

cure: something that makes an illness or injury better

hike: a walk in nature

improve: to get better

limb: a leg or arm

pace: rate or speed of movement

squeeze: to press something tightly together

venom: something an animal makes in its body that can harm other animals

FOR MORE INFORMATION

BOOKS

Cunningham, Malta. *How Do Snakes Poop?* North Mankato, MN: Capstone Press, 2019.

Gagne, Tammy. *Snakes: Built for the Hunt.* North Mankato, MN: Capstone Press, 2016.

Johnson, Rebecca. *Super Snakes.* New York, NY: Windmill Books, 2018.

WEBSITES

Snake
animals.sandiegozoo.org/animals/snake
Go to the San Diego Zoo's website to learn more about slithering snakes.

Snakes
www.dkfindout.com/us/animals-and-nature/reptiles/snakes/
Find out more about these remarkable reptiles here.

Super Snakes
kids.nationalgeographic.com/explore/nature/super-snakes/
Head to this website to read more fun facts about these awesome animals.

Publisher's note to educators and parents: Our editors have carefully reviewed these websites to ensure that they are suitable for students. Many websites change frequently, however, and we cannot guarantee that a site's future contents will continue to meet our high standards of quality and educational value. Be advised that students should be closely supervised whenever they access the internet.

INDEX